COMPETITIVE CHEMISTRY 2

INTRODUCTION

This objective chemistry series provides a basic and challenging problem of chemistry from particular topics. It can be used to brush up ones basics and checking up the preparation level of particular topic. It is equally helpful to the traditional classes as well as competitions. It can be also taken as a revision material for any competition which includes the test of basic chemistry. If you want to grasp the subject before practicing these multiple choice questions, you can go through the website http://www.ncert.nic.in/ncerts/textbook/textbook.htm and down load the free copy of science books and after having command on the topic practice it.

If you have any query or suggestion about this series you can send your suggestion at uk2594@gmail.com.

CONTENTS

CHAPTER 5   CHEMICAL REACTION AND EQUATIONS

CHAPTER 6   ACID BASES AND SALTS

CHAPTER 7   METALS AND NON METALS

CHAPTER 8   CARBON AND ITS COMPOUNDS

CHAPTER 9   PERIODIC CLASSIFICATION OF ELEMENTS

5. CHEMICAL REACTION AND EQUATIONS

SOME IMPORTANT POINTS

➢ Chemical reaction has taken place if

    1. Change in state

    2. Change in colour

    3. Evolution of gas

    4. Change in temperature

➢ The symbolic representation of chemical reaction is chemical equation
➢ Mass can neither be created nor be destroyed in a chemical reaction
➢ The number of atom of each element is equal on both side of equation such equation called balanced chemical equation
➢ Those reaction in which two or more substance combine to form a new substances called combination reaction
➢ A compound splits in two or more substances such reaction called decomposition reaction
➢ Reaction in which heat is absorb called endothermic reaction
➢ Reaction in which heat is released called exothermic reaction
➢ Decomposition reaction carried out by :

    1. Heat

    2. Electricity

    3. Sunlight

➢ Reaction in which one element takes place of another element in a compound are called displacement reaction
➢ Reaction in which two compounds react by the exchange of ions to form two new compound are called double displacement reaction
➢ Reaction in which addition of oxygen and loss of hydrogen in a substance oxidation reaction takes place

- ➢ Reaction in which addition of hydrogen and loss of oxygen in a substance reduction reaction takes place
- ➢ When a metal is attached by a substances it like moisture, air and water etc. And thi process is called corrosion
- ➢ Nitrogen gas is flushed in packets of chips to prevent the chips getting oxidised

5. CHEMICAL REACTION AND EQUATIONS

1. Which of the following is Chemical reaction?

   (a) Milk is left a room temperature during summer

   (b) food digested in our body

   (c) Ice melts

   (d) both (a) and (b)

2. What would be formed when magnesium:

   (a) magnesium oxide   (b) Magnesium dioxide

   (c) magneseoxide   (d) none of these

3. Which of the following called skeletal Chemical equation

   (a) Balanced equation   (b) unbalanced equation

   (c) both (a) & (b)   (d) None of these

4. $Zn + H_2SO_4 \longrightarrow$ _____ + _____

   (a) $Zn_2SO_4+H_2O$ (b) $ZnSO_4+H_2$

   (c) $ZnSO_4+H_2O$ (d) $Zn_2SO_4 + H_2$

5. $3Fe + 4H_2O \longrightarrow$ _____ + 4_____

   (a) $Fe_3O_4+H_2$ (b) $Fe_3O_4+H_2O$

   (c) $Fe_2O_3+H_2O$ (d) $Fe_2O_3 + H_2$

6. $CO + 2H_2$ _(340 atm)_ _____ + _____

   (a) $CH_3(OH)_3$ (b) CH_4 (c) CH_3OH (d) CH_2OH

7. $6CO + 6H_2O$ _sunlight, chlorophyll_ _____ + _____+$6O_2$

   (a) $C_{12}H_{22}O_{11}$ (b) $C_6H_{12}O_6$ (c) $C_6H_{10}O_7$ (d) None

8. Fe + HCl $\longrightarrow$ _____ + _____

   (a) $FeCl_3 + H_2$ (b) $FeCl_2 + H_2$ (c) $Fe(Cl_3)_2 + H_2$ (d) $FeCl_2 + H_2O$

9. $CaO + H_2O \longrightarrow$ _____ + _____

   (a) CaOH   (b) $CaOH_2$ (c) Ca_2OH (d) $Ca(OH)_2$

10. What is the chemical formula of marble?

    (a) $Ca(OH)_2$ (b) $CaCO_2$ (c) $CaCO_3$ (d) $CaCO_4$

11. Which of the followings are used for the white washing of the walls?

    (a) Calcium Carbonate   (b) Calcium Hydroxide

    (c) Calcium Bicarbonate   (d) All of these

12. Which of the following is an example of exothermic reaction?

    (a) $CH_4 + 2O_2 \longrightarrow CO_2 + H_2O$

    (b) $2AgBr \xrightarrow{sunlight} 2Ag + Br_2$

    (c) $H_2O \xrightarrow{electricity} H_2 + O_2$

    (d) $2Ba(OH)_2 + NH_4Cl \longrightarrow 2BaCl_2 + NH_4OH$

13. which of the following reaction is used in black and white photography.

    (a) $Pb(NO_3)_2 \xrightarrow{heat} PbO + 4NO_2$

    (b) $2AgCl \xrightarrow{sunlight} 2Ag + Cl_2$

    (c) $AgBr \xrightarrow{sunlight} 2Ag + Br_2$

    (d) Both b and c

14. What is the colour of ferrous sulphat?

    (a) Blue   (b) Brown   (c) Green   (d) Yellow

15. which of the following is an example of displacement reaction?

(a) Fe + $CuSO_4$ $\longrightarrow$ $FeSO_4$+ Cu

(b) Zn + $CuSO_4$ $\longrightarrow$ CO_2+ H_2O

(c) Mg + $CuSO_4$ $\longrightarrow$ $MgSO_4$+ Cu

(d) All of these

16. The given equation is the example of:

$CaO + H_2O \longrightarrow Ca(OH)_2$ (aq.)

(a) Combination (b) Decomposition

(c) Displacement (d) Double Displacement

17. Which of the following is an example of double displacement reaction?

(a) $FeSO_4$ $\longrightarrow$ Fe_2O_3 + SO_2 + SO_3

(b) Pb + $CuCl_2$ $\longrightarrow$ $2PbCl_2$ + Cu

(c) $Na_2SO_4 + BaCl_2$ $\longrightarrow$ $BaSO_4$ + NaCl

(d) None of these

18. $2Pb(NO_3)_2$ $\xrightarrow{\text{heat}}$ $2PbO + 4NO_2 + O_2$

The above reaction is an example of?

(a) Combination (b) Decomposition

(c) Displacement (d) Double Displacement

19. Which type of reactions are called exothermic reactions?

(a) in which heat is released (b) in which heat is absorbed

(c) Both a and b (d) None of these

20. The decomposition reaction is carried out by applying:

(a) light (b) heat (c) electricity (d) all of these

21. $Na_2SO_4 + BaCl_2 \longrightarrow BaSO_4 + 2NaCl$

In the above reaction, write the name of precipitate:

(a) Na_2SO_4 (b) $BaCl_2$ (c) $BaSO_4$ (d) NaCl

22. Which reaction is called oxidation reaction?

(a) in which oxygen is added   (b) in which hydrogen is removed

(c) Both a and b   (d) None of these

23. The reaction $2Cu + O_2 \xrightarrow{heat} 2CuO$

(a) Combination   (b) Oxidation   (c) Reduction   (d) Both a and b

24. What is the colour of the CuO in reaction?

$2Cu + O_2 \xrightarrow{heat} 2CuO$

(a) Blue   (b) Black   (c) Brown   (d) None of these

25. Which of the following is a example of redox reaction?

(a) $CuO + H_2 \xrightarrow{heat} Cu + H_2O$

(b) $ZnO + C \longrightarrow Zn + CO$

(c) Both a and b

(d) None of these

26. Which gives oxygen for oxidation or remove hydrogen is termed as ?

(a) Oxidizing agent   (b) Reducing agent

(c) Redoxing agent   (d) None of these

27. The black coating on silver is an example of:

(a) Rancidity   (b) Corrosion   (c) Reduction   (d) None of these

28. Which gas is used for protecting chips from rancidity?

(a) Argon   (b) Neon   (c) Hydrogen   (d) Nitrogen

29. Which reaction produces insoluble salts?

(a) Decomposition   (b) combination

(c) Precipitate   (d) Displacement

30. Which reaction occurs when an element displaces another element from its compound?

(a) Decomposition   (b) combination

(c) Precipitate   (d) Displacement

31. Which reaction is opposite of decomposition reaction?

(a) Combination   (b) Displacement   (c) Reduction   (d) Oxidation

32. Which of the following is an example of combination reaction?

(a) $Mg + O_2 \longrightarrow MgO$

(b) $NO + O_2 \longrightarrow NO_2$

(c) $H_2 + Cl_2 \longrightarrow HCl$

(d) All of these

33. Which of the following property a balanced chemical equation have?

a. Equal number of reactant and equal number of product.

b. Equal no. of atoms in reactants and products.

c. Different no. of atoms in reactants and products.

d. All of these

34. Which of the following is an oxidizing agent?

(a) $KMnO_4$ (b) $K_2Cr_2O_7$ (c) HNO_3 (d) All of these

35. $CuO + H_2 \longrightarrow Cu + H_2O$

In above reaction which substance is getting reduced?

(a) CuO   (b) Cu   (c) H_2O (d) H_2

36. Oxygen and moisture are necessary for?

(a) Corrosion   (b) Rusting of iron

(c) Both a and b   (d) None of these

37. A white coloured compound ……………..chloride changes grey when exposed to sun light.

(a) silver   (b) bromine   (c) barium   (d) none of these

38. $CaCO_3 \xrightarrow{\text{heat}} CaO + CO_2$

(a) Thermal decomposition reaction

(b) Electrolytic decomposition reaction

(c) Photolytic decomposition

(d) None of these

39. Complete the equation:

$Fe + HCl \longrightarrow$ _____ $+ H_2$

(a) $FeCl_2$ (b) $FeCl_3$ (c) Fe_2Cl_3 (d) Fe_3Cl_4

40. Burning of Coal is an example of?

(a) Exothermic reaction   (b) Endothermic reaction

(c) Decomposition   (d) none of these

41. Aluminium + sulphuric acid $\longrightarrow$ ………..+ ……………

(a) Aluminium Sulphate, Water   (b) Aluminium Sulphate, Hydrogen gas

(c) Aluminium Oxide, Water   (d) Aluminium Hydroxide, Hydrogen gas

42.   What is the colour of copper Sulphate?

(a) Blue   (b) Black

(c) Green   (d) Brown

43.   In which reaction exchange of ions occurs?

(a) Combination reaction   (b) Decomposition

(c) Displacement   (d) Displacement

44. $H_2S + Cl_2 \longrightarrow$ ....+...................

(a) SO_2 +HCl   (b) $SCl_2 + H_2$

(c) S + HCl   (d) S_2 + HCl

45.   Fe + S $\longrightarrow$ ....

(a) Fe_2S (b) FeS

(c) Fe_2S_3 (d) None of these

46. $Ca(OH)_2 + CO_2 \longrightarrow$ ....+ H_2O

(a) CaO   (b) $CaCO_2$

(c) $CaCO_3$ (d) $Ca(OH)_2$

47.   What is the formula of slaked lime.

(a) CaO   (b) $Ca(OH)_2$

(c) $CaCO_3$ (d) None of these

48. $2Al + 3CuCl_2 \longrightarrow$ ....+ ....

(a) 2AlCl, 3Cu   (b) $2AlCl_3$, 3Cu

(c) 3AlCl$_2$, 3Cu

(d) None of these

49. Zn + 2HCl $\longrightarrow$ ....+ ....

(a) ZnCl$_2$, H$_2$

(b) ZnCl, H$_2$

(c) ZnCl$_2$, H$_2$O

(d) ZnCl, H$_2$O

50. 3H$_2$ + N$_2$ $\longrightarrow$ ....

(a) 2NH$_3$

(b) 2NH$_4$

(c) 2N$_2$H$_3$

(d) 2NH$_2$

ANSWERS:

QUE.	ANS.	QUE.	ANS.	QUE.	ANS.	QUE.	ANS.	QUE.	ANS.
1	D	11	B	21	D	31	A	41	B
2	A	12	A	22	C	32	D	42	A
3	B	13	D	23	D	33	B	43	D
4	B	14	C	24	B	34	D	44	C
5	A	15	D	25	C	35	A	45	B
6	C	16	A	26	A	36	C	46	C
7	B	17	C	27	B	37	A	47	B
8	A	18	B	28	D	38	A	48	B
9	D	19	B	29	C	39	B	49	A
10	C	20	B	30	D	40	A	50	A

6. ACIDS, BASES AND SALTS

SOME IMPORTANT POINTS

- Acids are sour in taste change blue litmus into red
- Bases are bitten in taste and change red litmus into blue
- Turmeric and red cabbage ate natural indicator whereas Methyl orange and phenolphthalein ate synthetic indicator
- Vanilla, onion and clove ate olfactory indicator
- Acid + metal $\rightarrow$ salt + hydrogen gas
- Acid + metal carbonate / metal hydrogen carbonate $\rightarrow$ salt + carbon dioxide + water
- Acid + base $\rightarrow$ salt + water
- This reaction is known as neutralization reaction like a base
- Non- metallic oxides are acidic in nature and behave like acid
- Acid and base release its ions water
- Acid release H^t, H_3O^t ions
- Base release OH^- ions
- Acid and base in water conduct electricity
- Bases which are soluble in water called alkalis
- The mixing of acid and base in water this reaction is highly exothermic
- The PH of a neutral solution is 7 and PH of acid is less than 7 and of base is more then 7.
- Our body works in PH 7.0 to 7.8 PH of rain water is less than 5.6 it is called acid rain tooth decay begins when PH of mouth is lower than 5.5
- Bleaching powder chemical name – calcium oxy chloride " formula – caocl2 used in textile paper Indus try and disinfecting drinking water
- Baking soda – chemical name – sodium hydrogen carbonate " formula $\rightarrow$ NAHCO3
- Washing soda – chemical name – sodium carbonate deca hydrates formula – $Na_2Co_3. 10H_2O$
- Water of crystallization is the fixed number of water molecules present in one formula unit of a salt

➤ Plaster of parts

Chemical name – Calcium Sulphate hemihydrates

Formula – $CaSO_4 . 1/2H_2O$

➤ Gypsum

Chemical name – calcium sulphate dihydrate " formula – $CaSO_4 . 2H_2O$.

6. Acid, Bases and Salts

1. Acid is _____ in taste.

    a. Sweety   b. Bitter   c. Sour   d. None of these

2. Acid changes the colour?

    a. Litmus Blue to Red litmus   b. Litmus Red to Blue litmus

    c. Litmus Red to Yellow   d. All of these

3. A Strain of curry on a white shirt becomes reddish brown. when washed with deterg
    it becomes _____ in colour .

    a. Yellow   b. Red   c. Blue   d.Raddish

4. Bases are _____ in taste?

    a. Sour   b. Bitter   c. Sweet   d. None of these

5. Bases change the colour?

    a. Red litmus to Blue litmus   b. Blue litmus to Red litmus

    c. Radish Brown litmus to Red litmus   d. None of these

6. Which are following a synthetic indicator for acids and bases ?

    a. Phenolphthalein   b.Lemon   c. Methyl orange   d. Both(a)&(b)

7. When we react dilute sulphuric acid, added few pieces of zinc granules , then

    Pass the gas being evolved through the soap solutions them bubble will form

    , take burning candle near the bubble there would be pop sound so the gas

    Present the bubble is ?

    a. Oxygen   b. Hydrogen   c. Nitrogen   d. None of these

8. The metals combine with the remaining part of the acid and form a compound

is called ?

a. Base   b. Acid   c. Salt   d. None of these

9. The reaction of the acid with a metal can be summarised as?

a. Acid + Metal →Salt + Hydrogen gas

b. Acid + Salt → Metal + Hydrogen gas

c. Acid + Hydrogen gas → Metal + Salt

d. None of these

10. Limestone, chalk and marble are different from of ?

a. Sulphuric acid   b. Calcium carbonate   c. Hydrocarbonate   d. None of these

11. All metal carbonates and hydrogen carbonates react with _____ to give a

Corresponding salt, carbon dioxide and water

a. Acid   b. Salt   c. Bases   d. None of these

12. The reaction between an acid and a base to give a salt and water is known as?

a. Neutralisation   b. Oxidation   c. Addition   d. All of these

13. The general reaction between a metal oxide and an acid can be written as?

a. Base + Acid → Salt + Water   b. Metal oxide + Acid → Salt + Water

c. Acid + Salt → Water + Metal   d. All of these

14. What happen when small amount of copper oxide in a beaker and add dilute

Hydrochloride acid ?

a. The solution becomes blue – green   b. The metallic oxide change in gas

c. The copper oxide dissolve .   d. Both (a)&(c)

15. The reaction of a base with an acids metallic oxides are said to be?

a. Phynolic oxide   b. Acidic oxide   c.Basic oxide   d.All of these

16. When we touch the blue litmus the colour will change into?

    a. Yellow   b. Red   c. Green   d. Reddish blue

17. Acid found in?

    a. Lemon   b. Curd   c. Oranges   d. All of these

18. Which gas is usually liberated when an acid reacts with a metal?

    a. Hydrogen gas   b. Oxygen gas   c. Nitrogen gas   d. Helium gas

19. When we mix acid with base element remains?

    a. Salt   b. Base   c. Acid   d. Acid remains   d. All of these

20. Choose wrong statement?

    a. All acids have similar chemical properties

    b. All acids have different chemical properties

    c. Acids are sour in taste   d. Base change the colour of items red litmus

21. Which element is common for all acids?

    a. Oxygen   b. Hydrogen   c. Helium   d. Argon

22. Which of solution do not conduct electricity ?

    a. Glucose   b. Salty water   c. Alcohol   d. Both (a)&(c)

23. The electric current is carried through the solution by?

    a. Ions   b. Molecules   c. Isotopes   d. All of these

24. The cation present in acids is ?

    a. H^+ b. H^- c.E^- d. E^+

25. Acids produce hydrogen ions, H^+ (aq), in solution which have are responsible

For?

a. Basic property   b. Acidic property   c. Alcoholic property   d.None of these

26.   Which of the chemical is alkali?

a. Barium hydrochloride   b. Sodium hydrochloride

c. Both (b)&(d)   d. Calcium hydrochloride

27.   Choose the chemical formula of hydrochloric ion is?

a. H_4O^+ b. H_2O c. H^+O_4 d. H_3O

28.   Hydrogen ions cannot exist after alone, but they exist after combing with?

a. Air molecule   b. Hydrogen molecule

c. Water molecule   d. None of these

29.   Hydrogen ions cannot exist alone , but they exist after combing with water

molecules , Thus hydrogen ions must always be shown as $H^+(aq)$ or hydrogen

ion (H_3O^+) is summarise as ?

a. $H^+ + 2HO \rightarrow H_3O^+$ b. $H^+ + H_2O \rightarrow H_2O$

c. $H^+ + H_2O \rightarrow H_3O^+$ d. None of these

30.   Acids gives _____ or _____ in water.

a. H_3O^+ or $H^+(aq)$ b. H_2O or $H^+(aq)$ c. H_3O or H_2O d. None of these

31.   Bases generate _____ in water?

a. Hydrogen   b. Oxygen   c. Hydrochloride   d. None of these

32.   Which are soluble in water are called?

a. Alkalis   b. Acidic   c. Alkali   d. All of these

33.   The processing of dissolving an acid or a base in water is highly?

a. Exothermic   b. Intrinsic   c. Acidic   d. Basic

34.   Human stomach produces?

        a. Hydrochloric acid   b. Hydrochlorine   c. Bases   d. All of these

35.   To get rid of stomach pain, people uses bases called ?

        a. Hydrochloric acid   b. Sulphuric acid   c. Antacids   d. All of these

36.   The ph scale measure concentration of _____ in a solution ?

        a. Oxygen   b. Nitrogen   c. Helium   d. Hydrogen

37.   The ph scale of a neutral is?

        a. 6   b. 6.9   c. 7   d. 7.2

38.   Value less than 7 on the ph scale represents on ?

        a. Basic solution   b. hydrochloric solution

        c. Acidic solution   d. None of these

39.   The ph values less than 7 on the ph scale representation an increase in _____ in

        ion?

        a. OH^- b. E^- c. OH^+ d. All of these

40.   In the ph scale 7 to 14 represents _____ nature

        a. Acidic   b. Basic   c. Hydrochloric   d. None of these

41.   If we take hydrochloric acid of the same concentration, sexy one molar, then

        these produce _____ amount of hydrogen ions

        a. Same   b. Different   c. Big   d. None of these

42.   Acid that gives rise to more H^+ ions are said to be _____

        a. Strong acids   b. Weak acids   c. Acid in solid   d. All of these

43.   When ph of rain is less than _____ is called acid.

        a. 6   b. 7   c. 5.6   d. 8.2

44.   Human's tooth made up of hardest substance in the body is _____

        a. Calcium   b. Calcium   c. Calcium phosphate   d. None of these

45.   Stinging hair of nettle leaves inject _____ cause burning pain.

        a. Methanoic acid   b. Ethanoic acid   c. Base   d. None of these

46.   Acid present in curd is?

        a. Ethanoic acid   b. Citric acid   c. Acetic acid   d. Lactic acid

47.   Which of the following is the formula of salt?

        a. Nacl   b. Cl H_2O c. Na_2SO_4 d. Both(a) &(c)

48.   Salt of a strong acid and a strong base are with ph value of?

        a. 3   b. 7   c. 8   d. 5.5

49.   The common salt thus obtained is an important raw material for various materials of
        daily use, such as ?

        a. Baking soda   b. Sodium hydrochloride

        c.Bleaching powder   d. All of these

50.   When electricity is passed solution of sodium hydrochloride. This process

        called ?

        a. Chlor- Alkali   b. Chlor – Alkali   c. Aldehyde   d.None of these

51.   Bleaching powder is represent as ?

        a. 2NaoH(aq)   b. H_2(g)   c. 2NaCl(aq)   d. $CaoCl_2$

52.   Choose the correct use bleaching powder ?

a. For bleaching cotton and linen   b. As an oxidising agent in chemical industry

c. Potable water to make free of germ   d. All of above

53.   On heating gypsum at 373 k it losses water molecule and becomes calcium

Sulphate hemihydrates ($Caso_4$ ½ H_2O)this is called ?

a. Base   b. washing   c.Baking soda   d. Plaster of Paris

Answers :

Q	A	Q	A	Q	A	Q	A	Q	A	Q	A
1	C	10	B	19	A	28	C	37	C	46	D
2	A	11	A	20	B	29	C	38	C	47	C
3	A	12	A	21	B	30	A	39	A	48	B
4	B	13	B	22	D	31	C	40	B	49	D
5	A	14	D	23	A	32	A	41	B	50	A
6	D	15	C	24	A	33	B	42	A	51	D
7	B	16	B	25	B	34	A	43	C	52	D
8	C	17	D	26	C	35	C	44	B	53	D
9	A	18	A	27	D	36	D	45	A		

7. METAL AND NON METAL

SOME IMPORTANT POINTS

- Metals are lustrous, malleable, ductile, sonorous, hard, goods conductors where as non-metals are not
- Non-metals are brittle, generally low melting point
- Iodine is non-metal but it is lusturous
- Diamond is non-metal but it is hardest natural substance in the would
- Lead and mercury are metals bat poor conductors of heat and electricity
- Graphite is non-metal but it is conduct electricity
- Lithium, sodium, potassium can be cut by a knife
- metals have low ionization energy it is the energy is the amount of energy required to remove the electrons from the valance shell
- metal + oxides $\rightarrow$ metal oxide
- metal oxides react with acid as well as base are called amphoteric oxides
- aqua regia is freshly prepared mixture of concentrated hydrochloric acid and concentrated nitric acid in the ration of 3:1
- ionic compound are hard, soluble in water, rigid structure and conduct electricity in molten state
- the ore obtained after mining from the ground contain large amount of can they and sandy impurities called gangue
- cinnabar is ore of mercury
- calcination it is process of heating carbonate ore in lack of air
- roasting it is process of heating sulphide ore in excess of air
- thermit reaction used for welding of railway tracks
- galvianisation is a method of protecting steal and iron from rusting by coating then with a thin layer of zinc
- an alloy homogeneous mixture of two or more metal or a metal and metal
- solder is alloy of lead and tin has a low melting point used for welding electrical wires.

7. METAL AND NON METAL

1. Metals have shining surface. This property is called?

   a.Sonours   b.Hardness   c. Malleable   d. Metallic lusture

2. Metals can be beaten into thin sheets .This property is called _____?

   a.Sonours   b. Hardness   c.Malleable   d.Metallic lusture

3. Which of the following metals is most malleable metals?

   a.Gold   b.Silver   c. Lead   d.Both (a)&(b)

4. The ability of metals to be drawn into thin wire is called _____?

   a.Sonours   b.Ductility   c.Malleability   d.Regidity

5. 1 gm of gold can be drawn to from a wire of about _____ length.

   a.2000km   b.2000m   c.2000dm   d.None of these

6. Which metals are comparatively poor conductors of heat?

   a.Lead and Mercury   b.Mercury and silvers

   c.Mercury and copper   d.Lead and copper

7. The Wires that carry in your homes which material is coated on them?

   a.Polychloride   b.Polyvinychloride   c.Polyviny   d.None of these

8. Which of the following non metal is good conductor of electricity?

   a. Carbon   b.Iodine   c. Graphite   d.Diamond

9. Which of the following metal is a liquid at room temperature ?

   a. Mercury   b.Bromine   c.Xenon   d.None of these

10. Which metals have very low melting point gallium?

    a.Mercury   b.Caesium   c.Arsenic   d.Both(a)&(b)

11.   _____ is non metal but it is lustrous

a.Carbon   b.Iodine   c.Graphite   d.Radon

12.   Which metal have low density?

a.Lithium   b. Boron   c.Silicon   d.Aluminium

13.   Which oxides are produced when non metal are dissolved in water?

a.Basic   b.Acidic   c.Both(a)&(b)   d.None of these

14. $2Cu+O_2 \rightarrow$_____

a.CuO   b.2CuO   c.$2Cu_2O$ d.Cu_2O

15.   Which metals oxides which react with both acids as well as bases to produce

Salts and water?

a.Aluminium oxide   b.Zinc oxide   c.Both (a)&(b)   d.None of these

16. $Al_2O_3+2NaOH \rightarrow$ _____$+H_2O$

a. $2NaAlO_2$ b.$2NaAlO_3$ c.$2NaAlO$ d.$2Na_3Al_2O$

17.   Which of the following is correct chemical equation?

a.$Na_2O+H_2O \rightarrow NaH+O_2$ b.$Na_2O+H_2O \rightarrow NaOH$

c.$K_2O+H_2O \rightarrow K(OH)_2+H_2$ d.$K_2O+H_2O \rightarrow K(OH)_2$

18.   Which metal catch fire when kept in open?

a.K,Na   b.Na,I   c. Na,P   d. K,I

19.   Which Process is forming a thick layer of aluminium ?

a.Rusting   b. Anoding   c.Both (a)&(b)   d.Galvanisation

20. $Al_2O_3+6 HCL \rightarrow$ _____ + _____

a.$2AlCl_3 + 4H_2O$ b.$3AlCl + 3H_2O$ c.$2AlCl_3 + 3H_2O$ d. None of these

21. Which of the following react with cold water ?

a.Calcium   b. Magnesium   c.Zinc   d. Aluminium

22. $3Fe + 4H_2O \rightarrow$_____ $+ 4H_2$

a.Fe_2O_3 b.Fe_3O_4 c.Fe_4O_5 d.Fe O

23. Which of the following metal do not react with water at all ?

a. Iron   b. Aluminium   c. Sodium   d. Lead

24. Which of the following is more reactive ?

a. Iron   b. Aluminium   c. Sodium   d. Lead

25. Aquar regia is a freshly prepared mixture of ?

a.Dilute Hydrochloric acid and Dilute Nitric Acid

b.Concentrated Hydrochloric Acid and Concentrated Nitric Acid

c.Concentrated Hydrochloric Acid and Dilute Nitric Acid

d.Dilute Hydrochloric Acid and Concentrated Nitric Acid

26. Which of the following in correct for reactivity?

a.Al>Zn>Pb>Cu   b.Zn>Al>Pb>Cu   c.Al>Zn>Cu>Pb   d.Zn>Al>Cu>Pb

27. Which element has electronic configuration 2,8,8,1?

a. K-   b. Ar   c. K+   d.Ca+

28. Which of the following is a ionic compound?

a.N_2 b.NaCl   c.$MgCl_2$ d.Both (b)&(c)

29. Ionic compound soluble in ?

a.Kerosene   b.Petrol   c.Water   d.None of these

30. Which of the following element most aboudan on earth crust is _____.

a.Copper   b.Aluminium   c.Silver   d.Gold

31.   Which of the following element found in free state ?

a.Sodium   b.lead   c.Copper   d. Zinc

32.   The ore obtained after mining from the ground contains a large amount of

Earth and Sandy impurities .What they called?

O_2e b.Gangue   c.Mineral   d.None of these

33.   Cinna bar is an ore of _____ .

a.Mercury   b.Lead   c.Silver   d.Iron

34. $ZnCo_3$ connecting into oxide which process is used ?

a.Roasting   b.Calcination   c.Anodising   d.None of these

35. $3MnO_2 + 4Al \rightarrow 3$_____$+2$_____

a.$Al_2O_3 + Mn$ b.$AlO_3 + Mn_2$ c.$Mn + Al_2O_3$ d.$Mn_2 + AlO_3$

36.   Which of the following reaction in thermit reaction?

a.$3MnO_2 + Hal$ --------- _____ $3Mn + 2Al_2O_3$

b.$Fe_2O_3 + 2Al$ ---------- $2Fe + Al_2O_3$

c.$2Hgs + 3 O_2$ ---------- $2Hgo + 2So_2$

d.None of these

37.   The metal obtained is not pure is called _____

a.Curd Metal   b.Crude Metal   c.Curdy Metal   d. Surde Metal

38.   Whichof the following method used in ?

a. Galvanising   b.Annodising   c.Both(a)&(b)   d.None of these

39.   The alloy of measuring is _____.

a. Cinnabar   b.Amalgam   c.Solder   d.None of these

40. Which Metal oxide is used in preventing rusting of iron for iron pillar near Qutub minar, New Delhi?

a. Magnetic oxide   b.ZnO   c.Fe_2O_3 d.Al_2O_3

41. Steel is alloy of?

a. Iron , Carbon   b.Iron ,Silicon   c.Iron ,Nickel   d.None of these

42. Which alloy is used in welding of wires?

a. Amalgam   b.Solder   c.Steel   d.Cinnabar

43. Solder is an alloy of?

a. Lead ,copper   b.Copper,Tin   c.Lead,Tin   d.Tin,Copper

44. Which of the following alloy in rust free ?

a. Stainless Steel   b.Steel   c.Cinnabar   d.Amalgam

45. Bronze is an alloy of?

a.Copper and Zinc   b.Copper and Tin   c.Copper   d.Copper and Mercury

46. Iron,Nickel and Chromium are consistuent of which alloy ?

a.Stainless steel   b.Steel   c.Cinnabar   d.None of these

47. Which colour coating on copper carbonate when reacts with moist CO_2 ?

a.Blue   b.Brown   c.Green   d.Black

48. In thermit reaction which element powder is used as reducing agent?

a.Mn   b.Fe   c.Zn   d.Al

49. $Mg + 2HNO_3 \rightarrow$ _____ + _____

a.$MgNo_3 + H_2$ b.$MgNO_3 + H_2$ c.$MgCNO_3 + H_2$ d.$Mg(NO_3) + H_2O$

50. $ZnO + C \rightarrow$ _____ + _____

a. $Zn + Co_2$ b. $Zn + Co$ c. $Zn + Co_3$ d. None of these

Answers :

Q	A	Q	A	Q	A	Q	A	Q	A
1	D	11	B	21	A	31	A	41	A
2	C	12	A	22	B	32	B	42	B
3	D	13	B	23	D	33	A	43	C
4	B	14	B	24	C	34	B	44	A
5	B	15	C	25	B	35	C	45	B
6	A	16	A	26	A	36	B	46	A
7	B	17	B	27	D	37	B	47	C
8	C	18	A	28	D	38	C	48	D
9	A	19	B	29	C	39	B	49	C
10	D	20	C	30	B	40	A	50	B

8. CARBON AND ITS COMPOUND

SOME IMPORTANT POINTS

➢ The earth's crest has only 0.02% carbon in the form of mineral
➢ Most of carbon compounds are poor conductor of electricity and have low melting and boiling point
➢ The force of attraction between these molecules are not very strong
➢ Bonding in compounds are mainly covalent bond
➢ The two property of carbon tetra valency and catenation gives rise to large number of molecule
➢ The sharing of electron is known as covalent bond
➢ Carbon compound are two types saturated and unsaturated
➢ Saturated compounds has single bonds between the carbon atoms and unsaturated was double and triple bonds between carbon atoms
➢ Carbon compounds having identical molecular formula but different structure known as structural isomers
➢ The saturated hydrocarbons also called alkanes
➢ The carbon having double bond called alkene and triple bond called alkynes
➢ The formula of

$$Alkanes : C_NH_{2N+2}$$
$$Alkenes : C_NH_{2N}$$
$$Alkynes : C_NH_{2N-2}$$

➢ In hydrocarbons the element replaced the hydrogen is reffered as heteroatom
➢ These at heteroatom confer specific properties to the compound are called functional group
➢ A homologous series is a group of organic compounds having similar structure and similar chemical properties and successive compound is differ by CH_2 group
➢ When carbon compounds burnt in oxygen to give carbon dioxide along with the release of heat light

- Saturated hydrocarbons burns with a clean flame while unsaturated carbon compounds will give a yellow flam with lots of black smoke
- carbon compounds can be easily oxidized alcohols are converted to carboxylic acids by oxidation
- the industrial use of addition reaction in making vegetable
- the boiling of ethanol is 35k it is colourless liquid with pleasant small and it is miscible in water
- combustion of ethanol gives carbon dioxide and water + heat
- ethanol reacts with sodium to form sodium oxide and H_2
- the reaction between Ethanol and ethanoic acid is known as es trification and its reverse reaction is known as saponification
- the common name of ethanoic acid is acetic acid
- the dilute solution of acetic acid (5-8) in water is called vinegar
- the melting point of acetic acid is 190k and boiling is 391k
- ethanoic acid reacts with base to form salt and water
- it reacts with metal carbonates and hydrogen carbonates to form salt water and CO_2
- it is used in the form vinegar and as a food preservative for pickles sausages etc
- soaps are sodium or potassium salts of long chain of carboxylic acid
- detergents ate ammonium or sulphonate salts of long chain of carboxylic acid
- detergent is used in both hard and soft water but soap is used in only soft water and soap is bio-degradable

8. CARBON AND ITS COMPOUNDS

1. The earth's crust has carbon in the form of minerals of about:

   a) 5%   b) 3%   c) 0.02%   d) 0.03%

2. The ionic compounds have melting and boiling points at?

   a) High rate   b) low rate   c) Neutral   d) none of these

3. The melting point of Methane is:-

   a) 40k   b) 60k   c) 90k   d) 80k

4. The combining capacity of various elements dependss on the number of:-

   a) Protons   b) Valence electrons   c) Isotopes   d) None of these

5. The atomic number of carbon is:-

   a) 5   b) 4   c) 6   d) None of these

6. The reactivity of elements is explained as their tendency to attain a:-

   a) Noble gas Configuration   b) Ionic compound c)   Carbonic   Compound
   d) None of these

7. Carbon attains the noble gas configuration by

   a) Replacing electrons   b) Sharing electrons

   c)Graining electrons   d) None of these

8. How many electrons, helium has in its K shell

   a) 5 electrons   b) 4 electrons   c) 2 electrons
   d) None of these

9. How many electrons required by oxygen to complete its octet:-

   a) 2 electrons   b) 3electrons

   c) 8 electrons   d) none of these

10. The bond formed between two oxygen atoms is a:-

a) Single bond   b) double bond   c) Triple Bond
d) None of these

11.   A major Component of biogas is:-

a) Ethane   b) Pentane   c) Methane   d) None of these

12.   The word "Tetravalent" means:-

a) 4electrons in valence shell   b) 4 electrons in K Shell
c) 8 electrons in Valence shell   d) None of these

13.   The Covalent bonds are such bonds which are formed by:-

a) Sharing of electrons between two atoms

b) Isotopes

c) None of these

14.   Covalently bonded molecules have:-

a) High melting points   b) Low melting points

c) Neutral   d) None of these

15.   The atomic number of nitrogen is 7, its valency is:

a) 4   b) 5   c) 3   d) None of these

16.   Compound of carbon which is linked by single bonds is called:-

a) Saturated Compound   b) Unsaturated Compound
c) Flactuated Compound   d) None of these

17.   Compound of Carbon which is linked by double bonds is called:-

a) Saturated Compound   b) Unsaturated Compound
c) Flactuated Compound   d) None of these

18.   Compounds of carbon which are linked by triple bonds are called:-

a) Saturated Compound   b) Unsaturated Compound
c) Flactuated Compound   d) None of these

19.   The versatile nature of carbon consist

a) Tetravalency   b) Catenation   c) Both (a) & (b)   d) None of these

20.   Friedrich Wohler Prepare Urea by which compound:-

a) Ammonium Sulphate   b)AmmoniumCyanide
c) Sodium Carbonate   d) None of these

21.   The Saturated Compounds are normally:-

a) Highly Reactivity   b) Less reactive   c) Neutral   d) None of these

22.   The Structural Isomers Consists:-

a) Straight Carbon Chain   b) Branched Chain

c) Cyclic chains   d) All of these

23.   A Carbon Compound Methene is the member of

a) Alkane group   b) Alkene group   c) Alkyne Group   d) None of these

24.   A Compound of carbon, Haxane has the bond of:-

a) Single bond   b) Double bond   c) Triple bond   d) None of these

25.   The hydro Carbon compounds consists:-

a) Carbon   b) Hydrogen c) Both (a) & (b)   d) None of these

26.   The atom of element which replace the hydrogen atoms is called:-

a) Isotope   b) Heteroatom   c) Neutrons   d) None of these

27.   The first member of Alkane group is:-

a) Methene b) Ethane   c) Ethene   d) None of these

28.   The first member of Ketone group is:-

a) Methene b) Propanone   c) Butanone   d) None of these

29.   The first member of carboxyllicalid is:-

a) Methanal b) Ethanoicacid   c) Methanoic acid   d) None of these

30.   The isomer of Aldehyde group is:-

a) Ethanol   b) Ethanal   c) Propanone   d) None of these

31. Which is remain similar in a hhomologous series:-

a) Chemical Property   b) Melting Point   c) Static Store   d) None of these

32. The Chemical formula of Alkene is:-

a) C_nH_{2n+2} b) C_nH_{2n} c) C_nH_{2n-2} d) None of these

33. The Chemical formula of Alkane is:-

a) C_nH_{2n+2} b) C_nH_{2n} c) C_nH_{2n-2} d) None of these

34. The functional group of chloro-propane is:-

a) Aldehyde   b) Ketone   c) Halogen   d) Carboxyllicacid

35. Which suffix is present in members of ketone:-

a) -yne   b) -one   c) -oicacid   d) -al

36. When Carbon Compound, burn in air it gives:-

a) Carbon dioxide   b) Heat   c) Light   d) All of these

37. Which hydrocarbons when burn gives a yellow flame with lots of black smoke:-

a) Saturated hydrocarbons   b) Unsaturated Hydrocarbon

c) Complex Compounds   d) None of these

38. Which hydrocarbons give sooty flame when burning:-

a) Saturated hydrocarbons   b) Unsaturated Hydrocarbon

c) Complex Compounds   d) None of these

39. Which Catalyst is using in oxidizing reaction:-

a) Potassium   b) Alkaline Potassium Permanganate   c) Nickel   d) None of these

40. Vegetable oils generally have long chains of:-

a) Saturated Carbon   b) Unsaturated Carbon

c) Complex Carbon   d) None of these

41. Which catalyst is using in Addition reaction:-

a) Nickel   b) Ammonia   c) Copperd) None of these

42. What is added to iodine to make it tincture iodine:-

a) Carboxyllicacid b) Ethanol   c) Ammonium   d) None of these

43. Which is a physical property of Ethanol:-

a) Melting Point   b) Boiling Point   c) Good Solvent   d) All of these

44. Which plants are the convertors of sunlight into chemical energy:-

a) Wheat Plant   b) Cotton plant

c) Sugarcane Plant   d) None of these

45. When ethanol reacts with sodium it forms:-

a) Sodium Ethoxide   b) Sodium Ethanoate   c) Sodium hydroxide   d) All of these

46. When ethanoic acid reacts with sodium it forms:-

a) Sodium ethoxide   b) Sodium ethanoate

c) sodium hydroxide   d) All of these

47. Which is as Physical property of Ethanol acid:-

a) Sour tasteb) Melting Point   c) Boiling Point   d) All of these

48. Ethanoic acid reacts with ethanol to give:

a) Potassium   b) Ester   c) Sodium Saltd) None of these

49. Soaps are long chain carboxylic acid of:-

a) Ammonium Salts   b) Nickel Salts   c) Potassium Salts   d) None of these

50. The molecule of soaps consists:-

a) Hydrophilic ends   b) Hydrophobic ends

c) Both (a) & (b)   d) Sulphonate ends

Answers:

1. (c)	2. (a)	3. (c)	4. (b)	5. (c)	6. (a)	7. (b)	8. (c)	9. (a)	10. (b)
11. (c)	12. (a)	13. (a)	14. (b)	15. (c)	16. (a)	17. (b)	18. (b)	19. (c)	20. (b)
21. (b)	22. (d)	23. (b)	24. (a)	25. (c)	26. (b)	27. (c)	28. (b)	29.(c)	30. (b)
31. (a)	32. (b)	33. (a)	34. (c)	35 (b)	36. (d)	37. (b)	38. (a)	39. (b)	40. (b)
41. (a)	42. (b)	43. (d)	44. (c)	45. (a)	46. (b)	47. (d)	48. (b)	49. (c)	50. (c)

9. PERIODIC CLASSIFICATION OF ELEMENTS

SOME IMPORTANT POINTS

➤ Elements are classified on their similarities and differences.

➤ Dobereiner was the first scientist to show the relationship between the properties and their atomic masses.

➤ Law of triads states that when elements are arranged in order of increasing atomic masses in a group of three elements, the atomic, mass of middle element of the triads being equal to arithmetic mean of the atomic mass of the two elements.

➤ Newland's law of octaves states that when elements are arranged in order of increasing atomic masses, the properties of every eighth element is similar to the first element.

➤ Mendeleev arranged the element in order of increasing atomic mass and their chemical properties.

➤ Mendeleev left some spaces for few undiscovered elements in his periodic table.

➤ The anomalies of Mendeleev's periodic table are removed in Modern Periodic Table.

➤ In Modern Periodic Table going left to right in periods the atomic size and metallic properties decreases.

➤ In Modern periodic table going up to dawn in group, valency remains same, metallic properties and size of atom increases.

9. PERIODIC CLASSIFICATION OF ELEMENTS

1. In which year Doberenier made the triads

   a) 1917   b) 1817   c) 1816   d) 1815

2. Which of the following is not a traids

   a) Li, Na, K   b) Ca, Sr, Ba   c) Cl, Br, I   d) All of these

3. How many elements were found when Newland Classify elements:-

   a) 56   b) 55   c) 50   d) 57

4. Law of octane's applicable up to -----------

   a) Thorium   b) Calcium   c) Magnesium   d) Tungsten

5. How many elements were found in Mendeleev's time?

   a) 56   b) 63   c) 73   d) 53

6. Write the oxide of the tungsten.

   a) WO_2 b) WO_3 c) W_2O_3 d) W_7O_7

7. The Outer shells of the elements are completely filled are known as:-

   a) Nobel gas   b) Inert gas   c) Metalloid   d) both (a) & (b)

8. The atomic Mass of Silicon is :-

   a) 28   b) 28.1   c) 28.2   d) 28.3

9. The atomic No. Start from 58 to 71 is known as:-

   a) actinides   b) lanthanides   c) both (a) & (b)   d) none of these

10. The vertical columns are known as:

    a) Group   b) Rows   c) Period   d) both (a) and (b)

11. The elements belongs to the 17 group are known as:-

a) alkenes   b) actinides   c) Halogens d) both (a) & (b)

12. The atomic number of the elements is 14 is belong to the group in the periodic table.

a) 12   b) 13   c) 14   d) 9

13. The atomic radius is measured into

a) meter   b) Pm   c) um   d) hm

14. Atomic radius of Hydrogen atom is:

a) 37pm   b) 30pm   c) 37m   d) 30m

15. 1Pm = --------m

a) 10^{12}m   b) 10^{-12} c) 10^{-11} d) 10^{-10}

16. Which of the following is a non-metal

a) Iodine (I)   b) Hg   c) SC   d) Cs

17. Which of the following is a metalloid?

a) Re   b) Ir   c) Sb   d) Se

18. Modern periodic table is based on -------------

a) Atomic Mass   b) Increasing atomic number

c) Increasing atomic mass   d) Both (a) & (c)

19. Which of the followings is Reactive?

a) Rn   b) As   c) Xe   d) None of these

20. According to Mendeleev's periodic law, elements are arranged in

a) the increasing order of atomic mass

b) decreasing order of atomic mass

c) increasing order of atomic number

d) decreasing order of atomic number

21. Which of the following statement is correct about mendelliv's periodic law:-

    a) It has 18 Horizontal rows known as periods

    b) It has 6 horizontal rows know as periods

    c) It has 18 vertical columns know as group

    d) It has 7 vertical rows know as group

22. Which of the given elements A,B,C,D, and E with the atomic number 11,37,16,18, and 5 belong to the some group

    a) A,B,C   b) A,B   c) D,E,C   d) A,C,E

23. Where you would place element in the periodic table which electronic Configuration is 2,8,8:-

    a) 17th group   b) 18th group   c) 2nd group   d) 13 group

24. An element which is an essential constituent of all organic compounds belongs to

    a) 1st group and 2nd period   b) 14th group, 2nd period

    c) 14th group, 1st period   d) 13th group, 3rd period

25. Which of the following is the outer most shell for element of group 3?

    a) K Shell   b) M shell   c) N Shell   d) All of these

26. Which of the following elements exhibit maximum no. of valency electron.

    a) Cr   b) w   c) Zn   d) both (a) & (b)

27. Which of the following element exhibit maximum no. of Shells?

    a) Cr   b) Ne   c) Zn   d) Both (a) & (c)

28. Which of the following element has the largest number of radius?

    a) Po   b) As   c) O   d) W

29. Which among the following has the more metallic character?

    a) As   b) Be   c) Ba   d) both a and c

30. Among the following elements in the order of their decreasing atomic size Ag, Au, V, Ge, Ba is?

a) Ag>Ag>V>Ge>Be   b) Ge<V<Ag<Au<Ba   c)   Ba>Au>Ag>V>Ge
d) both (a) and (c)

31. What type of oxide would EKa-Boron forms?

a) BO_3 b) B_3O_2 c) B_2O_3 d) BO

32. These elements B, Si and Ge are:-

a) Metals   b) Non-Metals
c) Semi-Metal   d) Non-Metal, Metalloid and metals respectively

33. Which of the following element from an acidic oxide?

a) An element belongs to 15th group and 6th period

b) An element belongs to 14th group and 3rd period

c) An element belongs to 4th group and 4th period.   d) None of these

34. Which one of the following doesn't increase while moving down the group of the periodic table?

a) atomic radius   b) Metallic character   c) Valency   d) No. of shells

35. Write the Hydride forms by Eka-Sillico

a) SH_2 b) SH_4 c) S_2H_3 d) S_2H_5

36. The Atomic number gives us ------------------------

a) The Mass of Atom   b) No. of Protons present in the nucleus

c) No. of electrons   d) both (b) and (c)

37. How many noble gases discussed in the modern periodic table.

a) 7   b) 6   c) 5   d) 4

38. The electronic configuration of an element is 2,8,8,6

a) 5th period and 6th group

b) 4th period and 6th group

c) 4th period and 5th group

d) 3rd period and 3rd group

39. Which of the oxide forms by tungsten?

a) TgO_3

b) WO_3

c) W_2O_7

d) Tg_2O_7

40. The atomic radius of oxygen is:

a) 74pm   b) 63Pm   c) 66Pm   d) 72Pm

41. Which of these the atomic radius is largest

a) C   b) Li   c) Be   d) N

42. the atomic size refers to

a) Atomic Number

b) Atomic Mass

c) Atomic radius

d)atomic structure

43. Debierne's Classify only ----------- elements

a) 3   b) 6   c) 9   d) 12

44. Which of this element has no neutron is its nuclear:

a) Cl   b) As   c) H   d) He

45. The electronic configuration of H⁻ is

a) 2,1   b) 1   c) 2   d) 2,8,8,8,3

46. Which of this act like as metal as well as non metal?

a) Na   b) Cl   c) W   d) H

47. The atomic No. of Barium is

a) 37   b) 35   c) 56   d) 45

Answers:

QUE.	ANS.	QUE.	ANS.	QUE.	ANS.	QUE.	ANS.	QUE.	ANS.
1	B	11	C	21	B	31	C	41	B
2	D	12	C	22	B	32	C	42	C
3	A	13	B	23	B	33	B	43	C
4	B	14	A	24	B	34	C	44	C
5	B	15	B	25	D	35	D	45	C
6	B	16	A	26	C	36	D	46	D
7	D	17	C	27	D	37	B	47	C
8	B	18	B	28	D	38	B	48	
9	B	19	B	29	C	39	B	49	
10	A	20	A	30	B	40	C	50	

NOTES